YOUR KNOWLEDGE HAS VALUE

AF141038

- We will publish your bachelor's and
  master's thesis, essays and papers

- Your own eBook and book -
  sold worldwide in all relevant shops

- Earn money with each sale

Upload your text at www.GRIN.com
and publish for free

Bibliographic information published by the German National Library:

The German National Library lists this publication in the National Bibliography; detailed bibliographic data are available on the Internet at http://dnb.dnb.de .

Imprint:

Copyright © 2020 GRIN Verlag
Print and binding: Books on Demand GmbH, Norderstedt Germany
ISBN: 9783346141163

This book at GRIN:

https://www.grin.com/document/520517

Hemanta Kalita

Introduction to Core Modified Porphyrinoids. Smaragdy-rins, Expanded Porphyrinoids and Corroles

GRIN Verlag

GRIN - Your knowledge has value

Since its foundation in 1998, GRIN has specialized in publishing academic texts by students, college teachers and other academics as e-book and printed book. The website www.grin.com is an ideal platform for presenting term papers, final papers, scientific essays, dissertations and specialist books.

Visit us on the internet:

http://www.grin.com/

http://www.facebook.com/grincom

http://www.twitter.com/grin_com

Introduction to Core modified Porphyrinoids: Smaragdyrins, Expanded Porphyrinoids and Corroles

Dr. Hemanta Kalita

Department of Chemistry

IIT Bombay, Mumbai 400076

Table of Content

Chapter I: Smaragdyrins and Expanded porphyrinoids

1.1 Introduction

Porphyrins are a class of macrocycle with 18 π electron conjugated system consisting four heterocyclic rings connected by methine bridges with each other.[1] Porphyrins and metalloporphyrins present at the active site of numerous biomolecules play an important role in living systems.[2] Corroles are porphyrin analogues, tetrapyrrolic in nature, aromatic, and possesses contracted cavity due to absence of one *meso* bridged carbon atom connecting the heterocycles. In nature, porphyrins occur in various forms such as chlorin in chlorophyll pigments; as prosthetic heme group surrounded by the protein units in hemoglobin; as corrin in vitamin B_{12} etc.[3] Some of the notable biological roles served by porphyrin and its metal derivatives include electron transfer in cytochromes, radical generation in photosynthesis, oxygen storage and transport in myoglobin and hemoglobin in living organisms.[4] In addition, porphyrins and related macrocycles provide an extremely versatile synthetic base for a variety of materials applications.[5] Porphyrins and metalloporphyrins have broad range of applications as field-responsive materials, particularly for optoelectronic applications.[5]

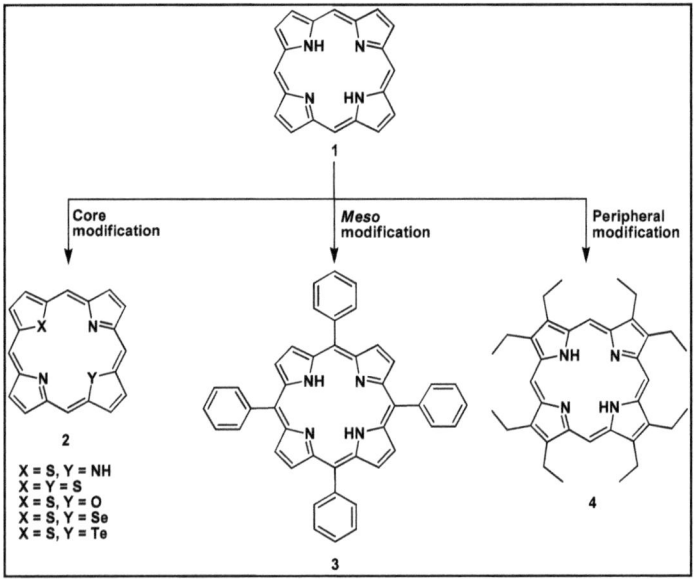

Chart 1.1.

4

The potential applications in the field of molecular electronics, catalysis, dye-sensitized solar cell, contrast agents for magnetic resonance imaging etc. caught chemists' attention to study these macrocyclic systems in depth. The diversity of porphyrin research has been broadened with the structural modification in various ways such as core modification, *meso*-modification, and peripheral β-substitution (Chart 1.1). Core modification involves substitution of one or two inner core pyrrolic nitrogen atom(s) by other atoms such as O, S, Se, Te, C, P, and Si.[6] The β-modification involves in substitution at the peripheral position of the heterocyclic rings. *Meso*-modification involves substitution at the *meso* position by aryl groups or by functional group(s). These structural modifications involved at the core, periphery and *meso* position leads to significant alteration in its electronic as well as structural properties.

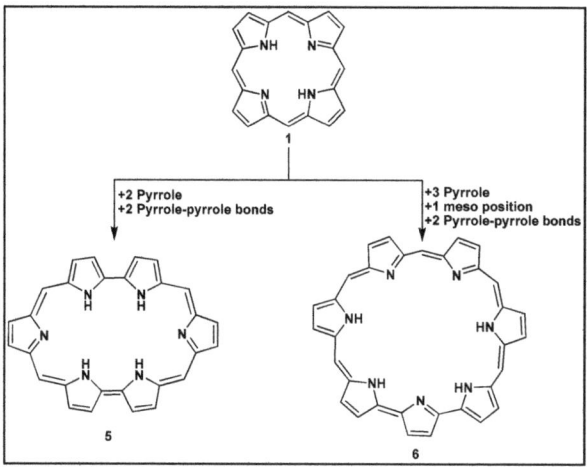

Chart 1.2.

Expanded porphyrins and expanded core-modified porphyrins are synthetic analogues of porphyrins and core-modified porphyrins respectively, containing more than 18π-electrons in a conjugated pathway either due to an increased number of heterocyclic rings or due to multiple *meso* carbon bridges. Expanded porphyrins containing a larger central core possess novel spectral and electronic features; interesting and often unprecedented cation-coordination properties; anion binding and in favorable cases anion transport; as sensitizers for Photodynamic therapy (PDT); as contrasting agents in magnetic resonance imaging (MRI); as radiation therapy enhancer; as nonlinear optical materials etc.[7] The number of π-electrons in the ring can be increased either by increasing the number of bridging atom(s) connecting the heterocycles or by increasing the number of heterocyclic rings (Chart 1.2). The expanded porphyrins such as 22π sapphyrins and smaragdyrins, 26π rubyrins, 30π heptaphyrins, 34π octaphyrins and higher cyclic polypyrrole analogues containing 40π, 48π, 64π, 80π and 96π systems had been reported in the literature. These macrocycles show rich structural diversity where normal and different kinds of inverted structures have been identified. There are rich literatures available on innumerous expanded porphyrins and related macrocycles. In this chapter, our focus is to discuss on smaragdyrins and its structurally related macrocyclic systems and their metal and nonmetal complexes.

1.2 Smaragdyrin

Smaragdyrin is a congener of expanded porphyrins family consisting of five heterocyclic rings. There are three different expanded porphyrins containing five heterocyclic rings such as pentaphyrin (7), sapphyrin (8) and smaragdyrin (9) (Chart 1.3). All these macrocycles are aromatic 22 π electron conjugated system, which follow Huckel's (4n+2) π electron rule. The difference among these macrocyclic systems lies in the number of bridging carbon atoms and the number of direct bonds that connect the heterocycles. Pentaphyrin contains five methine bridges; sapphyrin contains four methine bridges with one direct bond, and smaragdyrin contains three methine bridges and two direct bonds connecting the five heterocyclic rings. In other words, smaragdyrin is an expanded corrole, bearing three methine bridges and two direct pyrrole-pyrrole linkage. Sapphyrin was first reported by R. B. Woodward and co-workers at an aromatic conference in 1966.[8] In the same conference, Woodward et al. mentioned the possibility of existence of another macrocycle with five heterocyclic rings with three methine carbons and two direct bonds.[8] This macrocycle was named as smaragdyrin by Woodward because of its intense green colour. Unlike the chemistry of sapphyrin, the chemistry of smaragdyrin is still at the developing stage due to its inherent unstable nature and easily inaccessible precursors to synthesize these macrocycles.

6

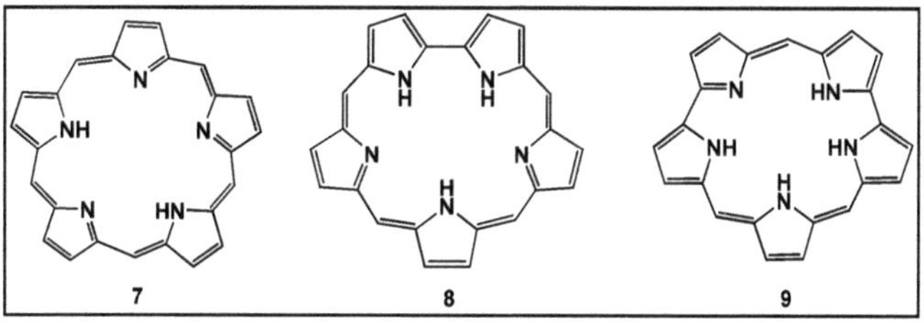

Chart 1.3.

1.3 Synthesis and characterization

A perusal of literature revealed that there are three main research groups who put effort to synthesize and stabilize the smaragdyrin macrocycle. In 1972 Broadhurst et al., realized the possibility of existence of another pentaphyrin macrocycle with two direct bonds during their rational synthetic approach of sapphyrin macrocycle.[9] These macrocycles called as norsapphyrins were synthesized in three steps as shown in Scheme 1.1.[9] These macrocyclic compounds are aromatic in nature and was confirmed by ^{1}H NMR which shows inner NH protons in high field region (4.85 ppm) and the *meso*-protons in down field region (~10.00 ppm) due to diamagnetic ring current effect. Dioxasmaragdyrin **16** was isolated in decent yield but their attempts to isolate its aza analogue following the same synthetic approach remained futile due to its highly unstable nature. The dioxasmaragdyrin showed several absorption bands in the region of 350-750 nm regions. The protonated form of

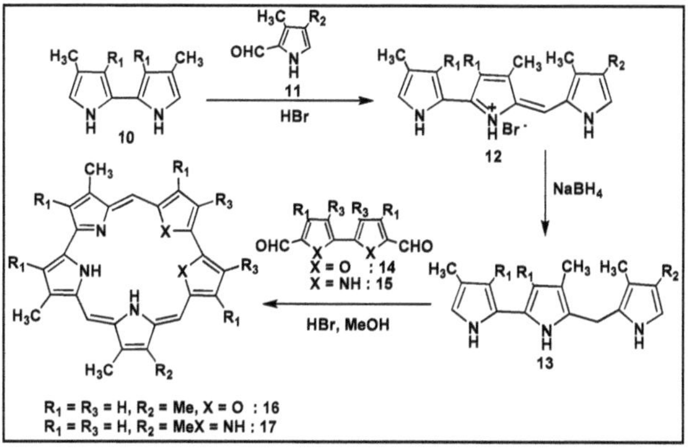

Scheme 1.1. Synthesis of norsapphyrins **16** and **17**.

7

dioxasmaragdyrin showed split Soret like band in addition to Q-bands. Owing to its inherent unstable nature and synthetic difficulties, the chemistry of this macrocycle remained unexplored for a long time.

In 1998, Sessler and co-workers, reported the synthesis of dibutyloctamethyl derivative of isosmaragdyrin **24** and its monoxa analogue **25**.[10] The isosmaragdyrin and its monooxa analogue were synthesized via acid-catalyzed condensation of bisformyl dipyrromethane **23** with **21** and **22** respectively, in 40% yield (Scheme 1.2).[10] The isosmaragdyrin derivatives are aromatic in nature as evidenced by NMR and absorption spectroscopy. The absorption spectra of compounds **24** and **25** showed split Soret band along with other Q-bands and the intensity of the Soret band of **25** is three times more than **24**. The authors succeeded in obtaining the single crystal X-ray structures of the hydrochloride salts of **24** and **25** (Figure 1.1a, 1.1b). The crystal structure revealed that the macrocyclic ring was nearly planar but the central pyrrole was deviated from the mean plane defined by all other nitrogen atoms by 23.2° in compound **24**. Except the nitrogen of the tilted pyrrole, all other four pyrrole

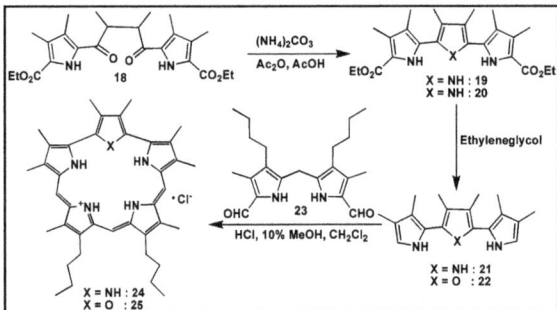

Scheme 1.2. Synthesis of compounds **24** and **25**.

nitrogens are bonded to the central Cl⁻ ion. Furthermore, the Cl⁻ ion was also found to be 1.919 Å away from the mean plane of the macrocycle. The X-ray structure of **25** exhibited similar features like the structure of **24**. The macrocycle **25** is relatively planar with the exception of central furan which is deviated away from the neighbouring pyrroles by 21.2 Å. The four pyrrole nitrogen atoms bound to Cl⁻ ion whereas furan oxygen was not involved

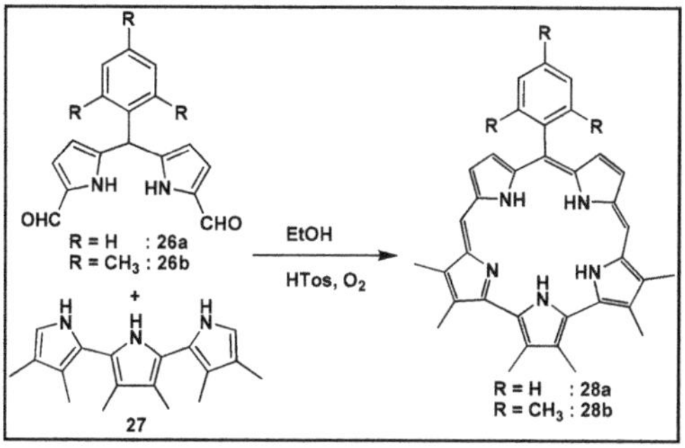

Scheme 1.3. Synthesis of azasmaragdyrins **28**.

in binding. The same research group also attempted to synthesize the partially β-substituted azasmargdyrin but were unsuccessful in isolating the same compound. They synthesized the azasmaragdyrin derivative **28a** by treating 1,9-bisformyl-5-pheyldipyrromethane **26** with hexamethylterpyrrole **27** in presence of HCl in chloroform, and isolated the desired the macrocycle as its hydrochloride salt (Scheme 1.3).[11] The X-ray structure of **28a** (Figure 1.1d) showed that the macrocycle is planar except for the deviation of middle pyrrole ring of terpyrrole moiety from the macrocyclic plane. The crystal structure revealed that the Cl⁻ ion was above the macrocyclic plane and bound to three nitrogen atoms.

Recently, Chandrashekar et al. were successful in preparing the first examples of stable *meso*-aryl core-modified smaragdyrin macrocyle by following Mc-Donald type [3+2] condensation approach.[12] They reacted 16-oxatripyrranes **30** and 16-thiatripyrrane **31** with *meso*-aryl dipyrromethanes **29** in the presence of acid catalyst TFA (Scheme 1.4), followed by aerobic oxidation with DDQ.[12] This strategy gave oxa or thia smaragdyrin derivative as the major product along with minor amount of corresponding oxa- or thiacorroles. However, the yield of the product by following this strategy strongly depends on the concentration of the acid catalyst used, *meso* substituent on the dipyrromethane moiety and the type of heteroatom present in the tripyrrane moiety. A large number of symmetric and unsymmetrical *meso*-aryl oxasmaragdyrins derivative prepared by following this synthetic route.[12]

9

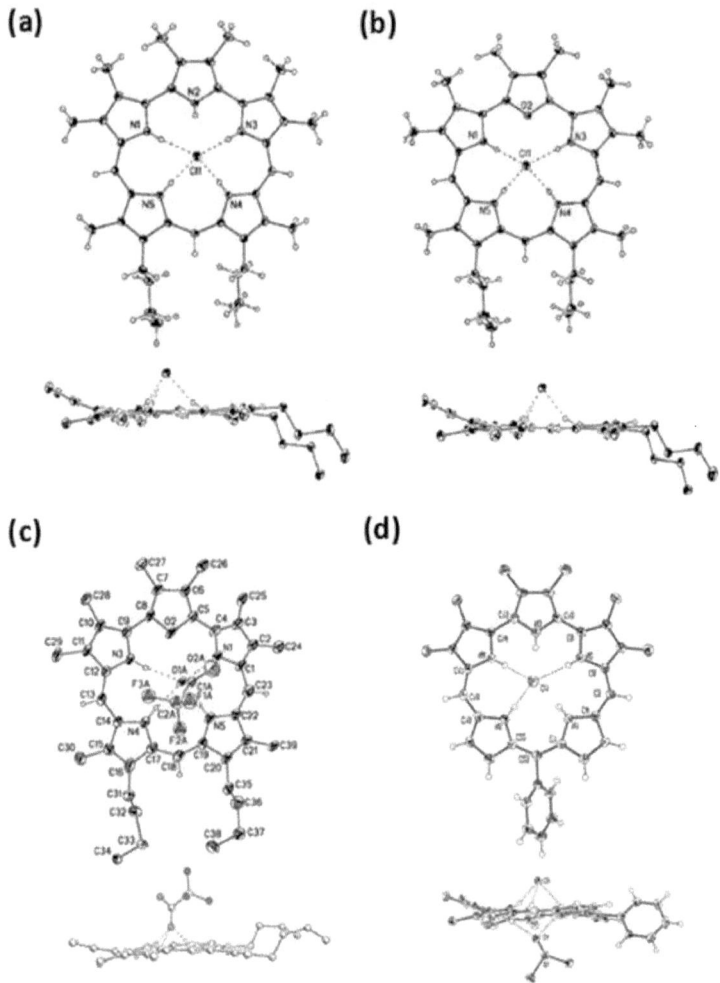

Figure 1.1. X-ray crystal structure of compounds (a) **24**, (b), (c) **25**, (d) **28a**. [From Pareek, Y.; Ravikanth, M.; Chandrashekar T.K. *Acc. Chem. Res.* **2012**, *45, 10,* 1801-1816. Reproduced with permission.]

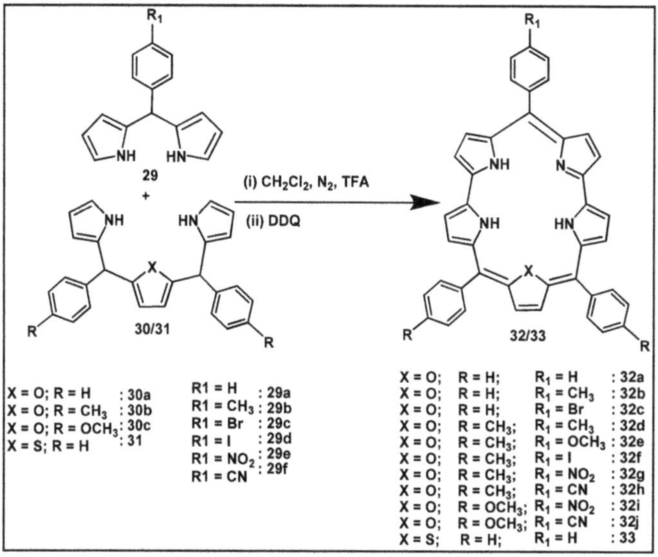

Scheme 1.4. Synthesis of *meso* aryl core modified smaragdyrins.

Although, the oxasmaragdyrin derivatives are prepared and studied extensively, the chemistry of thia analogue[12b] is limited, to one report, due to its poor yield and stability. However, the pentaza analogue of smaragdyrin cannot be prepared by this [3+2] synthetic route developed by Chandrashekar et al. The ^{1}H NMR of oxasmaragdyrin **32** exhibit four sets of doublet for β-pyrrole protons and one singlet for furan protons in 8.3-9.5 ppm region revealing the aromatic character of smaragdyrin macrocycle. The three inner NH protons which are expected to appear in high field region due to macrocyclic ring current were not observed even at -50 °C because of rapid tautomerism. The crystal structure showed that the macrocycle is not planar due to the strain caused by direct pyrrole-pyrrole link and steric repulsion among the amino hydrogen atoms (Figure 1.2).

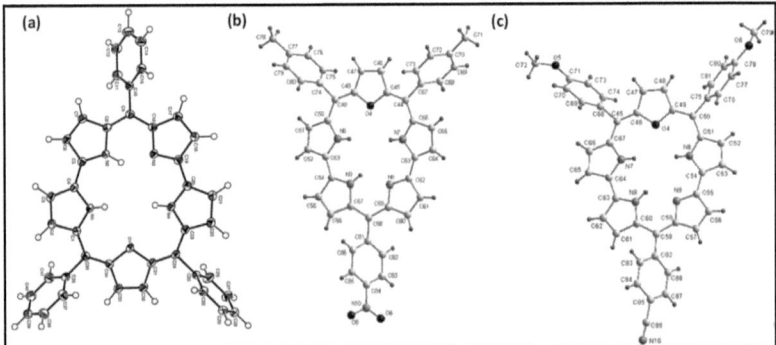

Figure 1.2. Crystal structure of (a) **32a**, (b) **32g** and (c) **32j**. [From Pareek, Y.; Ravikanth, M.; Chandrashekar T.K. *Acc. Chem. Res.* **2012**, *45, 10,* 1801-1816. Reproduced with permission.]

1.4 Metal and Non-metal Complexes of Smaragdyrin

Although smaragdyrin has a larger inner cavity core than regular porphyrin and more number of donor atoms available towards coordination, there are few reports on metal and non-metal complexes. Chandrasekhar et al. reported the Ni (34) Rh (35) complexes (Scheme 1.5) of *meso*-triaryl 25-oxasmaragdyrin.[12b] The coordination of Rh with smaragdyrin macrocycle involves the bonding in η^2-fashion with the dipyrromethene moiety with one amine and one imine nitrogen atom. The other two amine nitrogen atoms of the tripyrrane moiety were not involved in bonding with Rh (I). The crystallographic studies revealed that the Rh(I) formed an out-of-plane square planar complex with oxasmaragdyrin macrocycle. However, in case of nickel, although the crystal structure was not known, and they formulated as a π cation radical of Ni (II) based on paramagnetic nature of the complex as supported by NMR and EPR studies.

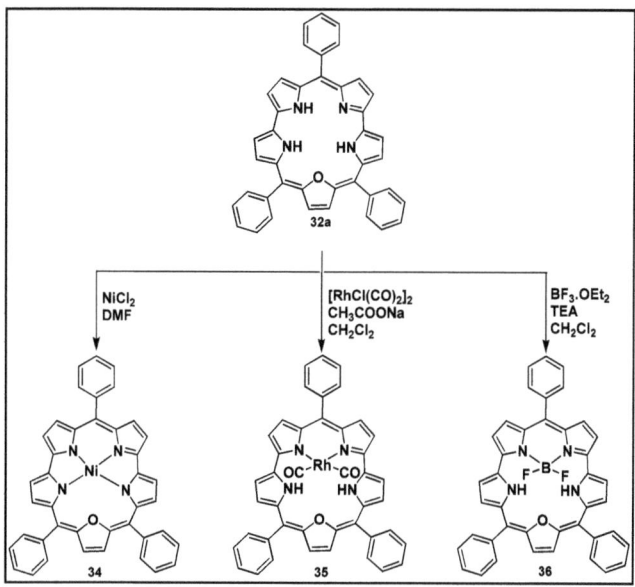

Scheme 1.5. Synthesis of metal and non-metal complexes of 25-oxasmaragdyrin.

Sessler and co-workers, reported mono-and bis-BF2 complexes of expanded porphyrins called amethyrin and [32]octaphyrin along with octaethylporphyrin by treating these macrocycles with BF3.OEt2 and NEt3 and characterized them by crystallography.[13] They found that the bis-BF2 complexes of octaethylporphyrin was not stable and undergoes rapid hydrolysis to form B-O-B bond. However, the bis-BF2 complexes of amethyrin and [32] octaphyrin are stable. Rao and Ravikanth synthesized the first B (III) complexes of triaryl 25- oxasmaragdyrin (36) bearing two fluorides as axial groups (Scheme 1.5), by treating oxasmaragdyrin with BF3.OEt2 and NEt3 in CH2Cl2 solvent at room temperature.[14] These BF2-smaragdyrin complexes are stable and does not undergo hydrolysis unlike previously reported BF2-complexes of octaethylporphyrin. They also prepared the B(OH)2-smaragdyrin derivative by treating the BF2-smaragdyrin with AlCl3 and water and showed its potential application of sensing fluoride ion.[14]

1.5 Properties of Smaragdyrin

The electronic spectra of oxa and thiasmaragdyrin exhibited a porphyrinoid-type Soret band ~445 nm and four Q-bands in the range of 500-730 nm.[12b] All absorption bands were found to be red shifted relative to corresponding N_3O corrole and porphyrin which is in agreement with the extension of conjugation in thia and oxa smaragdyrin. It is observed that thiasmargdyrin shows more red shifted absorption bands compared to the corresponding oxa derivative and it is probably due to the presence of bulky sulphur atom resulting in large distortion.[12b] The protonation results in splitting of Soret bands in both cases and causes a substantial red shift of all the bands in both oxa and thiasmaragdyrin. The metal derivatives also show similar absorption pattern like the protonated species of oxa and thiasmaragdyrin. The splitting of Soret and Q-bands of protonated and metal derivatives of smaragdyrin are in general agreement with the behaviour of *meso*-aryl expanded porphyrin and the splitting of Soret band is indicative of lowering of symmetry due to distortion as supported by their crystal structure.[12b] The BF_2-complex of oxasmaragdyrin showed two well-separated bands in the rangae of 445-475 nm regions like the protonated and metal derivatives along with Q-bands. The peak maxima of BF_2-complex of oxasmaragdyrin are slightly red-shifted compared to its free base. In addition to that, the extinction coefficients of BF_2-derivative are quite higher than its free base smaragdyrin. Interestingly, the absorption band ~700 nm is more intense by an order of three as compared to free-base smaragdyrin. The *meso*-aryl smaragdyrins are fluorescent and shows one emission band ~705 nm with a quantum yield of 0.04 with respect to H_2TTP.[14] On the other hand, the BF_2-smaragdyrin complex is more fluorescent with a single band emission that is bathochromically shifted by 10-15 nm from the last absorption band, with emission maximum centered around ~715 nm and the quantum yield is twice as high as compared to free base triaryl oxasmaragdyrin.[14] *Meso*-triaryl oxasmaragdyrin exhibits two reversible oxidations and two ill-defined irreversible reductions, which indicates that the macrocycle is electron-rich and stable under oxidation conditions as compared to free-base porphyrins. The BF_2-complex of *meso*-triaryl smaragdyrin shows two reversible oxidations and one reversible reduction and one irreversible reduction. The redox properties indicate that BF_2-complexation makes the macrocycle electron deficient, consequently oxidation is difficult and reduction is easier as compared to free base smaragdyrin under redox condition.[14]

14

1.6 Smaragdyrin Based Dyads and Triad

Although there are extensive literatures available on multiporphyrin arrays[15], to study energy transfer process and use as a functional materials, the reports on multi-expanded porphyrin arrays and porphyrin-expanded porphyrin arrays are still very few.[16] Chandrashekar et al.[14] reported the first *meso* free 25-oxasmaragdyrin and exploited the reactivity of free *meso* position to synthesize the first example of *meso-meso* linked smaragdyrin dyads 32 by treating *meso* free smaragdyrin monomers 37 either with AgPF6 or with *n*-BuLi (Scheme 1.6).[17] On the basis of theoretical calculations, authors proposed a twist angle of 71° to 64° between the smaragdyrin units in dyads 38 which resulted in enhanced conjugation leading to a red shift of the absorption spectra by 25 nm compared to the smaragdyrin monomer 37. The fluorescence emission spectra of smaragdyrin dimer exhibited slightly broadened emission as compared to monomer with peak maximum ~760 nm. The fluorescence quantum yields of monomer and dimer were found to be 0.042 and 0.018 respectively relative to tetraphenylporphyrin in benzene solvent.

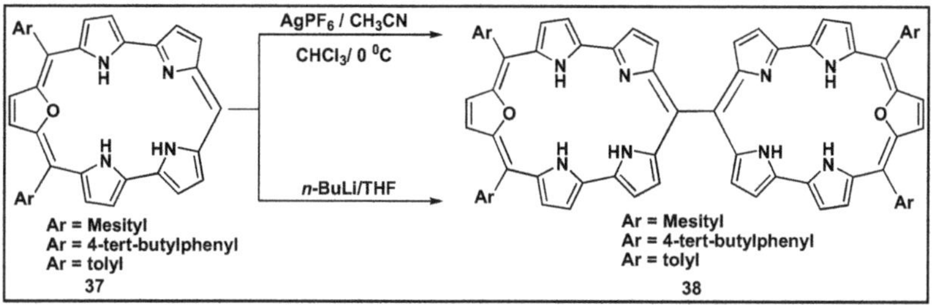

Scheme 1.6. Synthesis of *meso-meso* linked smaragdyrin dyads.

Ravikanth and co-workers reported the synthesis of diphenyl ethyne bridged porphyrin-smargdyrin dyad[18] (39), BF2-smaragdyrin-sapphyrin dyad[19] (40), and BF2-smaragdyrin-rubyrin[19] (41) dyad by coupling appropriate iodo functionalized and ethynyl functionalized porphyrin, expanded porphyrin monomeric building blocks under Sonogashira reaction conditions (Chart 1.4). The fluorescence studies carried out on dyad 39 supported an efficient singlet to singlet energy transfer from porphyrin unit to oxasmaragdyrin unit. The protonated 39 binds anion at the

15

protonated smaragdyrin site, can be seen from the gradual enhancement of porphyrin fluorescence indicating dyad **39** can be used as a fluorescent sensor for anions.

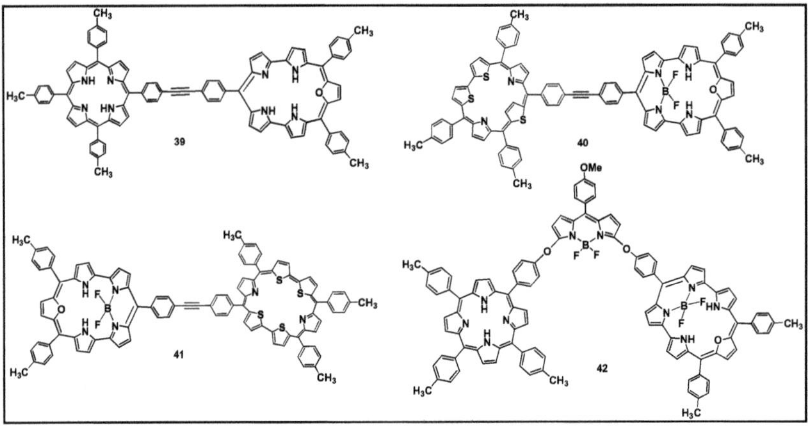

Chart 1.4.

The dyads **40** and **41** possess some special characteristics which make them favourable to use for fluorescent sensor applications.[19] These features are as follows:

(a) The expanded macrocycles thaiasapphyrin and thiarubyrin are non fluoresecent, but the BF$_2$-smaragdyrin macrocycle is fluorescent.

(b) Protonated form of sapphyrin and rubyrin can bind anions because of their large macrocyclic core cavity size as well as the availability of inner pyrrole NH protons.

(c) BF$_2$-smaragdyrin cannot be protonated and bind anions because the pyrrolic NHs are involved in hydrogen bonding with fluorides of the BF$_2$-group. Thus, the sensing studies on the protonated form of dyads **40** and **41** with various concentrations of CO$_3^{2-}$ ions indicated that there is a gradual enhancement of the fluorescent intensity of BF$_2$-smaragdyrin unit, where the BF$_2$-smargdyrin unit is acting as a signalling unit upon binding of anion at protonated sapphyrin/rubyrin site.

16

Khan and Ravikanth reported the trichromophoric triad **42** consisting of porphyrins, BODIPY, and BF₂-smargdyrin units and studied photophysical properties (Chart 1.4).[20] The fluorescence study of the triad **42** at the excitation wavelength of 488 nm, where BODIPY unit absorbs strongly, indicated that the emission from the BODIPY and porphyrin units were quenched by 99% and the major emission was observed from the BF₂-smaragdyrin unit. This is because the singlet state energy level of the BF₂-sargdyrin unit is lower than those of the other two chromophoric units and the energy transfer occurs efficiently from BODIPY and porphyrins units to BF₂-smaragdyrin unit in triad **42**.

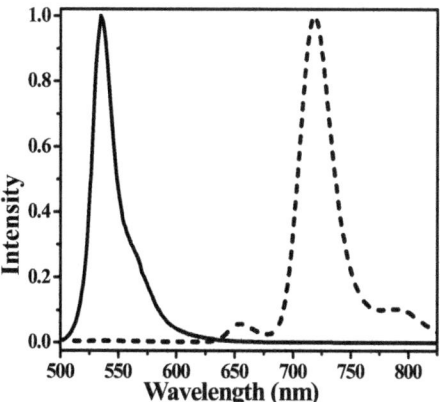

Figure 1.3. Emission spectra of triad **42** along with its constituted monomers exciting at 488 nm. The dashed line indicates the emission profile of the triad whereas the solid line indicates the emission spectrum of a 1:1:1 mixture of constituted monomers exciting at 488 nm. [From Khan, T. K.; Ravikanth, M. *Eur. J. Org. Chem.* **2011**, 7011-7022. Reproduced with permission.]

1.7 Phosphorus Complexes of Corroles and Expanded Macrocycles

The chemistry of phosphorus complexes of porphyrins[21], corroles[22], and N-fused porphyrins[20] are well known in the literature and a large number of literature reports are available. The insertion of phosphorus into these tetrapyrrolic macrocycles was carried out by refluxing free base macrocycle in pyridine with different phosphorylating reagents such as PCl₃, PCl₅, and POCl₃.[21-23] However, POCl₃ is found to be more suitable for phosphorus insertion in

17

these kinds of tetrapyrrolic macrocycles in recent times. The binding of phosphorus to N4 porphyrins and N4 corroles involves all the donor nitrogen atoms in the bonding along with one or two axial group(s) (Chart 1.5). In N4 porphyrins, phosphorus binds to four pyrrolic nitrogen atoms and two axial groups like chloro or hydroxo or one oxo group. Similarly, in N4 corrole, all the inner core nitrogen atoms are involved in bonding along with one or two axial group(s). In most cases the axial groups used to be one chloro or hydroxyl or one oxo group. The phosphorus atom in the central cavity is found to be hexacoordianted as well as pentacoordinated. All these phosphorus based macrocycles are important in various applications such as bio-imaging, catalysis etc.[22c, 22g]

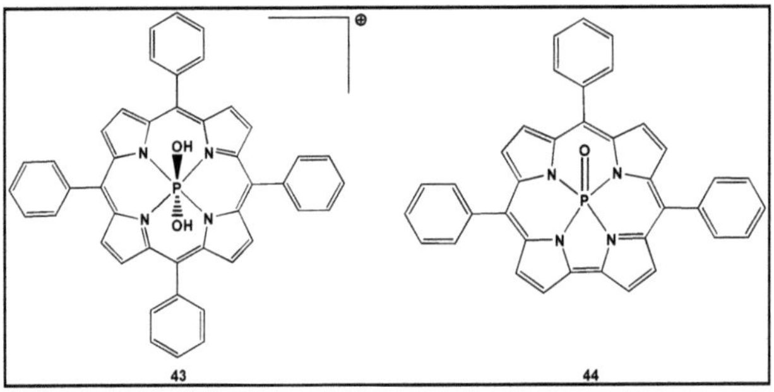

Chart 1.5.

However, phosphorus being a main group element, there is an upsurge interest in the study of phosphorus complexes of expanded porphyrin systems in recent times. It is noteworthy to mention that there are a few reports on phosphorus complexes of expanded porphyrinoids to understand the coordinating ability of phosphorus to expanded macrocycles and their properties and application. Osuka and co-workers synthesized the first examples of expanded isophlorins by treating octaphyrin **45** with 20 equivalents of PCl3 in the presence of triethylamine at room temperature and afforded monophosphorus **46** and bisphosphorus octaphyrin **47** complexes (Scheme 1.7).[24] In monophosphorus octaphyrin complex **46**, the phosphorus (V) was in trigonal bipyramidal geometry and bound to three pyrrolic nitrogen atoms and two pyrrolic β-carbon atoms. However, in bisphosphorus complex of isophlorin **47**, one phosphorus attains same trigonal bipyramidal geometry noted for monophosphorus complex but the second

18

phosphorus(V) was tetracoordinated and bound to two pyrrolic nitrogens, one β-carbon atom of pyrrole and the fourth position was occupied by one oxo group. Thus, phosphorus (V) takes two different coordination geometries in bisphosphorus complex **47**. Osuka and co-workers also prepared monophosphorus complex of

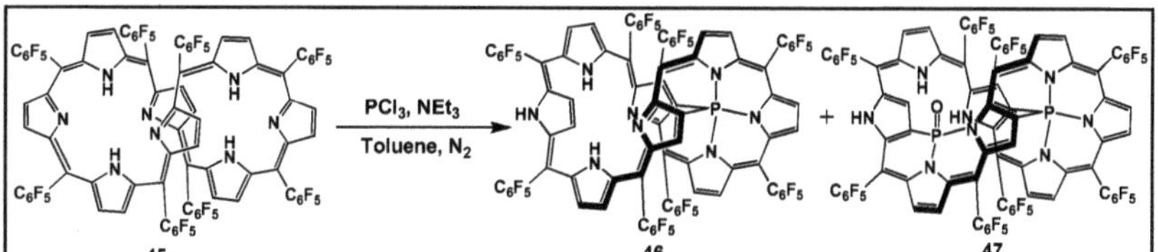

Scheme 1.8. Synthesis of monophosphorus and bisphosphorus octaphyrin complexes.

[28]hexaphyrin **49** and bisphosphorus complex of [30]hexaphyrin **50** and discussed their aromatic/anti-aromatic characteristic features (Scheme 1.8).[25] In [28]hexaphyrin **48**, the phosphorus (V) was bound to two pyrrolic nitrogen atoms, one β-carbon of pyrrole and one oxo group whereas in bisphosphorus complex of [30]hexaphyrin **50**, one phosphorus was in the same coordination environment as in [28]hexaphyrin whereas the second phosphorus(V) was in different coordination environment and bound to three pyrrolic nitrogen atoms and

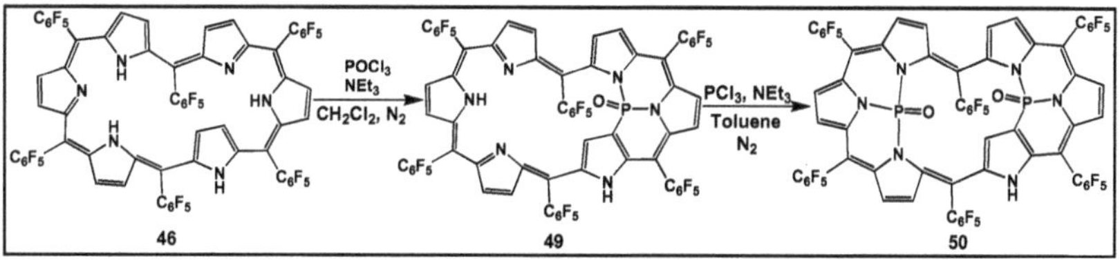

Scheme 1.9. Phosphorus complexes of hexaphyrin.

one oxo group. The absorption spectrum of **49** exhibits one Sore-like band ~588 nm and broad Q-like bands at 876 and 1055 nm. Moreover, compound **49**, shows an emission band at ~1090 nm which is a mirror image of the corresponding absorption spectrum. On the other hand, compound **50** shows ill-defined Soret-like bands and broad, weak NIR bands. In addition to that, compound **50** shows no fluorescence emission band. The same research group also reported the phosphorus insertion into N-fused [24]pentaphyrin (Scheme 1.10) which resulted in double fusion reactions and obtained phosphorus complex of a triply-fused [24]pentaphyrin **52**.[26] In triply fused [24]pentaphyrin **52**, the phosphorus atom was bound

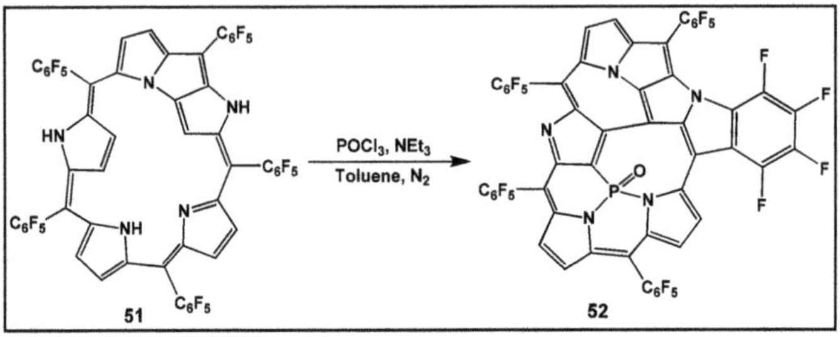

Scheme 1.10. Phosphorus complexes of N-fused pentaphyrin.

to two pyrrolic nitrogen atoms, one pyrrolic β-carbon and one oxygen atom and found to be antiaromatic. The absorption spectrum shows a blue-shifted Soret-like band and almost no Q-bands compared to N-fused [24] pentaphyrin **51**. However, the binding of phosphorus to *meso*-pentafluorophenyl-substituted [32]heptaphyrin is slightly different from that of the other expanded porphyrinoid. It involves the binding of four pyrrolic nitrogen atoms along with ortho carbon atom of one of the *meso*-pentafluorophenyl groups in a trigonal bipyramidal geometry.[27] The binding of phosphorus in N-fused porphyrin involves similar type as that of expanded porphyrinoid (Scheme 1.11).[23]

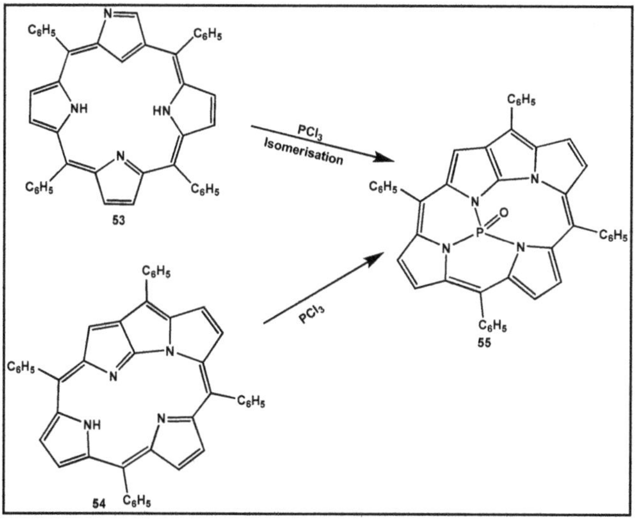

Scheme 1.11. Phosphorus insertion N-fused porphyrin.

Chapter II: Core-modified Corroles

2.1 Introduction

Corroles are tetrapyrrolic contracted porphyrins, fully conjugated 18π electron macrocyles containing three methine bridges and one direct pyrrole-pyrrole linkage.[28] Corroles owe their name to cobalt-chelating corrin of vitamin B_{12} with identical skeleton.[29] The 18 π system in contracted framework of corroles is maintained by the change in the oxidation state of one of nitrogen atoms.[30] Thus in corrole macrocycle **57**, there are three amine type nitrogen atoms and one imine type nitrogen atom in the central cavity, unlike porphyrins **56** having two nitrogen atoms of each type (Chart 1.6). As a result of lacking one methine carbon atom in corrole, the inner core of the macrocycle is smaller trianionic N_4-donors as compared to the dianionic N_4-donating porphyrins.

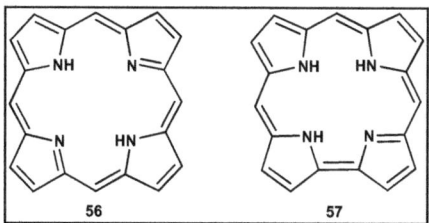

Chart 1.6.

Johnson and Kay[31] first reported corrole in 1964 and the first X-ray structural determination of a free-base corrole was done by Hodgkin and co-workers[32] in 1971.The same research group later reported the *meso*oxa and thiacorroles from the respective heteroatom containing precursors.[33] Although the synthesis of corrole had been reported long back, the synthesis of corrole remained difficult until 1999. Gross et al.[34] and Paolesse et al.[35] independently published the facile methodologies for the synthesis of corroles by commercially available precursors. There is vast literature available about N_4-corroles and its metal complexes. However, here we limit our discussion to core modified corroles.

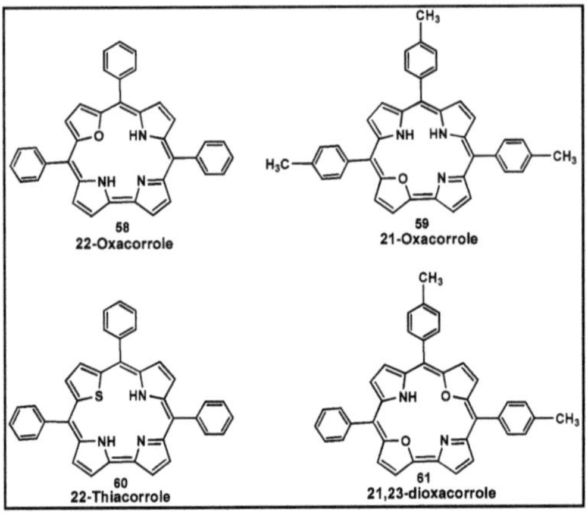

Chart 1.7.

2.2 Synthesis of Core-modified corrole

Core modified corroles are resulted from the replacement of one or two inner nitrogen atoms with other heteroatoms such as sulphur, oxygen, selenium, phosphorus, tellurium etc (Chart 1.7).[36] The synthesis of core modified corroles containing oxygen and sulphur was first reported by Broadhurst et al. but the synthesis involved inaccessible precursors.[37] Because of synthetic difficulties, the chemistry of core modified corroles had not been developed to greater extent unlike azacorroles.[38] In 1999, Chandrashekar et al. isolated the 22-oxacorrole as a side product during their successful synthesis of stable *meso*-triaryl smaragdyrins by [3+2] condensation approach.[12a] In this method, aryl oxasmargdyrin **32a** was the main product of the reaction whereas aryl 22-oxacorrole **58** was the minor product (Scheme 1.12). Later, in 2002, Lee et al. developed a method for the synthesis of triaryl oxacorrole which involved a [2+2] condensation approach (Scheme 1.13).[39] In this approach, the acid-catalyzed condensation of furylpyrromethane alcohol **62** with *meso*-aryldipyrromethane **29b** gave aryl 22-oxacorrole **64** in 15% yield whereas the condensation of regioisomer of furylpyrromethane alcohol **63** in the presence of acid and followed by

24

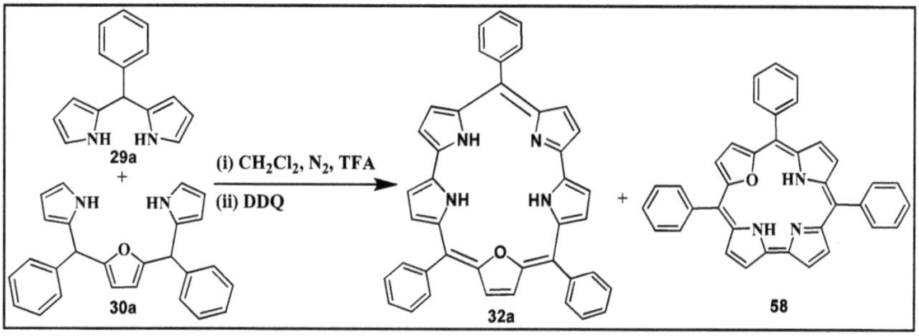

Scheme 1.12. Synthesis of 22-oxacorrole as a side product during oxasmaragdyrin synthesis.

oxidation with DDQ gave aryl 21-oxacorrole in 9% yield. Latos- Grazynski and co-workers reported the synthesis of 21,23-dioxacorrole (Chart 1.7) containing a direct pyrrole-furan link by acid-catalyzed condensation of 2,5-bis(p-tolylhydroxymethyl) furan, 2-(phenylhydroxymethyl)furan, and pyrrole in 1:1:2 molar ratio.[40] Shetti and Ravikanth recently reported the aryl 22-thiacorrole (Chart 1.7) which involved the propionic acid reflux conditions.[41]

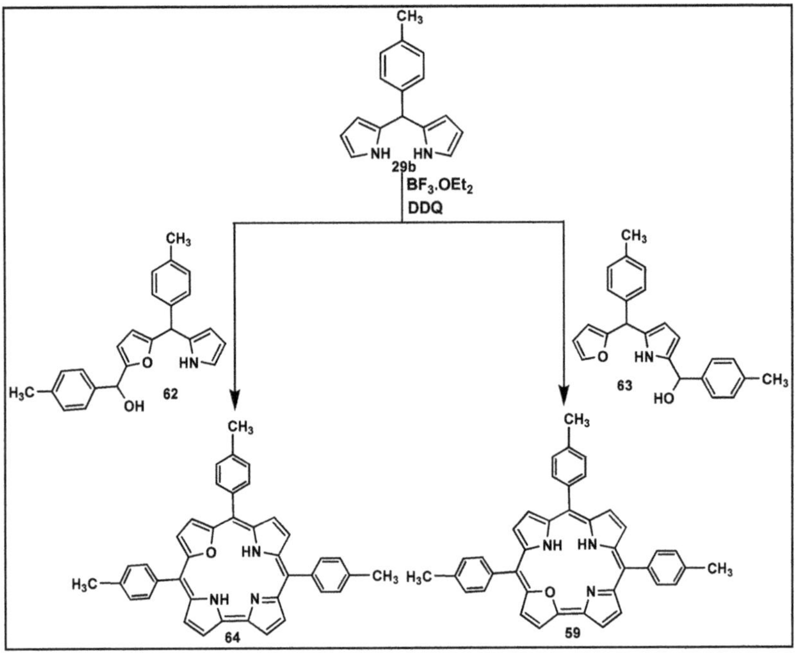

Scheme 1.13. Synthetic strategy for 21-, and 22-oxacorrole followed by Lee et al.

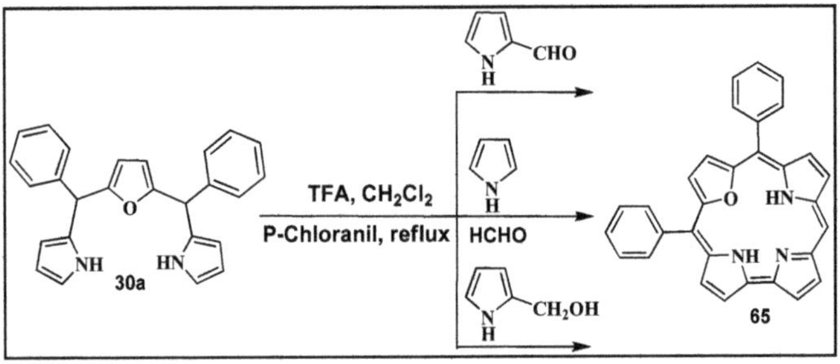

Scheme 1.14. Synthesis of *meso*-free 22-oxacorrole.

To synthesize *meso* free 22-oxacorrole, Chandrashekar et al. used three different synthetic routes.[42,43] As discussed earlier, they obtained they obtained *meso*-aryl 22-oxacorrole as minor product by condensing 16-oxatripyrrane and *meso*-aryl dipyrromethane.[12a] To synthesize *meso* free 22-oxacorrole, they revamped their strategy by condensing the pyrrole-2-carboxaldehyde with 16-oxatripyrrane in the presence of 1eq. TFA in CH_2Cl_2 followed by oxidation with *p*-chloranil (Scheme 1.14). The second method involved the condensation of 16-oxatripyrrane with pyrrole and paraformaldehyde in the presence of TFA. This strategy worked out well when a slight higher concentration of acid was used. In the third method, 2-hydorxymethyl pyrrole was condensed with 16-oxatripyrrane to afford *meso* free 22-oxacorrole **65** using the similar reaction conditions. The concentration of acid catalyst plays a pivotal role in the formation of 22-oxacorrole.

Although the inner core of monooxacorrole is similar to porphyrin core, dianionic in nature, there are few metal complexes reported in literature unlike porphyrins.[44] This is possibly because the core ring is contracted in 22-oxacorrole due to presence of one direct bond. Chandrahekar and co-workers reported the Rh(I), Cu(II), Co(II) and Ni(II) complexes of triaryl oxacorrole.[44] In Rh (I) complex, one amino and one imino nitrogen atoms were coordinated to the metal to form metal complex. All the inner core heteroatoms were coordinated to the metal atom in case of other three metal complexes.

2.3 Spectral Properties of Corroles

The triphenyl 22-oxacorrole show one Soret type intense band ~410 nm and four Q-type bands in the region of 490-640 nm regions.[44] All these Soret and Q-type bands are hypsochromic shifted compared to tetraphenylporphyrin (H_2TTP). This is due to replacement of one nitrogen atom by oxygen in the central core. Similar observation is found on going from H_2TTP to monooxa-tetraphenyl porphyrins. Upon protonation with TFA, the Soret band is split, which is due to lowering of symmetry in solution. Spectroscopic properties suggest that oxacorrole molecule is aromatic. Triphenyl monooxacorrole shows very strong emission with a quantum yield of 0.88 relative to H_2TTP.[42] The emission maxima are strongly blue shifted relative to the corresponding porphyrin. Protonation leads to the red shift of emission band and quantum yield of 0.23 of the protonated derivative, which is still higher than the corresponding porphyrin. The *meso* free diphenyl 22-oxacorrole show similar absorpting properties, one Soret band ~405 nm and four Q-type bands in the region of 490-625 nm regions. The absorption bands are blue shifted by 5-15 nm compared to its triphenyl analog. Emission properties show that the *meso* free diphenyl corrole showed one emission band ~ 650 nm and one shoulder ~700 nm. The quantum yield is as low as four times (0.19) relative to triphenyl 22-oxacorrole.

2.4 Corrole Based Dyads

The studies of extended conjugated macrocycles such as porphyrin, corrole are of considerable interest because of the potential applications such as optoelectronic materials, sensitizers in biomedical processes, and contrasting agents in magnetic resonance imaging (MRI). The main advantage of these extended macrocycles is large delocalized π-electronic conjugation and their stability under ambient conditions required for such applications. The extension in conjugation results in the reduction of frontier oribtals energy gap, thereby altering the energies of the electronic tranisitions, and shift of the absorption maxima towards red region.[45] Thus, these macrocycles can be used readily as biochemical sensitizers as their absorption maxima fall within the therapeutic window of 600-800 nm. There are few methods reported in literature to extend the π-delocalization pathway of porphyrins. One of these synthetic stratagies involve increase in the number of heterocyclic rings or the number of *meso* links, that are connected to the pyrrolic rings, resulting in the formation of expanded porphyrins. The other alternate strategy to obtain larger framework of π electrons is to connect individual porphyrin, expanded porphyrins or BODIPY units to one another to make discrete arrays of controllable length. Osuka and co-workers, synthesized a large multiporphyrin array by linking the *meso* position of one porphyrin unit with other

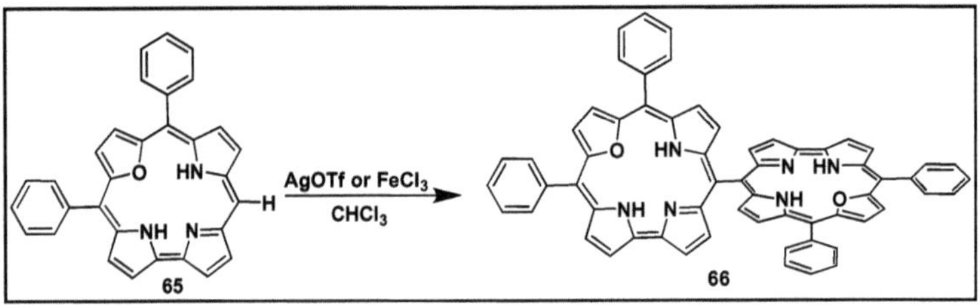

Scheme 1.15. Synthesis of *meso-meso* 22-oxacorrole dyad.

28

porphyrins regioselectively through silver-salt coupling methodology.[46] This resulted in the formation of covalently linked porphyrins oligomers with interesting physical and electronic properties. If the singly linked porphyrins arrays are further coupled with suitable reagents *meso-meso* and *β-β* doubly and triply linked porphyrins tapes are formed. The electronic interaction between each porphyrin unit makes it possible to alter the frontier orbitals energy gap to a considerable extent. These molecules are planar, and absorbs in the near-infrared (NIR) region because of their hugely extended π delocalization. However, there are few reports on the use of the structurally similar 18π corroles as promising electronic materials. The recent synthetic advancement in corrole chemistry led chemist to synthesize *meso* free monoxa corrole in good yields. The *meso*-free position of monooxa corrole is reactive like porpyrin and when it was treated with AgOTf or FeCl₃ under ambient conditions, the *meso-meso* linked dimer **66** is formed in quantitative yields (Scheme 1.15). This was the first example of *meso-meso* linked dimer in corrole chemistry reported till date[47] in contrast to a *β-β* linked dicorrole, as reported by Gross et al[48]. The nonlinear optical (NLO) properties of the oxacorrole monomer and dimer were studied by Rayleigh scattering method and found a moderated increase in the *β* values of the dimers relative to those of monomers indicated the effect of extension of π-conjugation. Furthermore, the Cu (II) metalated dimer was found to be as a potential DNA-cleaving agent upon exposure to visible light. The absorption spectrum of such dimer showed that there is a ~10 nm red shift in the peak maxima, and this small shift is attributed to the weak interaction of the mommeric units.

References

(1) Lindsey, J. S. in *The Porphyrin Handbook* (Eds), Kadish, K. M.; Smith, K. M.; Guilard. R. Academic Press: San Diego, CA, **2000**, *Vol. 1*, p.45-118.

(2) (a) Dunford, H. B. In *Peroxides in Chemistry and Biology*; Everse, J.; Everse, K. E.; Grisham, M. B., Eds.; CRC Press: Boca Raton, FL, **1991**; Vol. 2 p 1. (b) Kadish, K. M.; Smith, K. M.; Guilard, R. *The Porphyrin Handbook*, Vol. 1-20, Academic Press: San Diego, CA, **2000**.

(3) (a) Deisenhofer, J.; Epp, O.; Miki, K.; Huber, R.; Michel, H. *Nature* **1985**, *318*, 618.   (b) Deisenhofer, J.; Epp, O.; Miki, K.; Huber, R.; Michel, H. *J. Mol. Biol.* **1984**, *180*, 385.

(c) Takano, T. *J. Mol. Biol.* **1977**, *110*, 537.   (d) Geno, M. K.; Halpern, J. *J. Am. Chem. Soc.* **1987**, *109*, 1238.

(4) (a) Dawson, J. H.; Sono, M. *Chem. Rev.* **1987**, *87*, 1255. (b) Zinth, W.; Knapp, E. W.; Fischer, S. F.; Kaiser, W.; Deisenhofer, J.; Michel, H. *Chem. Phys. Lett.* **1985**, *119*, 1. (c) Tabushi, I.; Kugimiya, S-i.; Kinnaird, M. G.; Sasaki, T. *J. Am. Chem. Soc.* **1985**, *107*, 4192. (d) Collman, J. P.; Wagenknecht, P. S.; Hutchison, J. E. *Angew. Chem. Int. Ed.* **1994**, *33*, 1537.

(5) (a) Holm, R. H. *Chem. Rev.* **1987**, *87*, 1401. (b) Mlodnicka, T. In *Metalloporphyrins in Catalytic Oxidation*; Sheldon, R. A., Ed.; Marcel Dekker: New York, 1994, p. 275. (c) Menieur, B. *Chem. Rev.* **1992**, *92*, 1411. (d) Dolphin, D.; Traylor, T. G.; Xie, L. Y. *Acc. Chem. Res.* **1997**, *30*, 251. (e) Ni, C.-L.; Abdalmuhdi, I.; Chang, C. K.; Anson, F. C. *J. Phys. Chem.* **1987**, *91*, 1158. (f) Collman, J. P.; Hutchison, J. E.; Lopez, M. A.; Tabard, A.; Guilard, R.; Seok, W. K.; Ibers, J. A.; L'Her, M. *J. Am. Chem. Soc.* **1992**, *114*, 9689. (g) Collman, J. P.; Ha, Y.; Wagenknecht, P. S.; Lopez, M.-A.; Guilard, R. *J. Am. Chem. Soc.* **1993**, *115*, 9080. (h) de Rege, P. J. F.; Williams, S. A.; Therien, M. J. *Science* **1995**, *269*, 1409. (i) Drain, C. M.; Fischer, R.; Nolen, E. G.; Lehn, J.-M. *J. Chem. Soc., Chem. Commun.* **1993**, 243. (j) Funatsu, K.; Kimura, A.; Imamura, T.; Ichimura, A.; Sasaki, Y. *Inorg. Chem.* **1997**, *36*, 1625. (k) Wojaczynski, J.; Latos-Grazynski, L. *Inorg. Chem.* **1995**, *34*, 1044.

(6) Gupta, I.; Ravikanth, M. *Coord. Chem. Rev.* **2006**, *250*, 468.

(7) (a) Pandey, R. K.; Zheng, G. *Porphyrins as Photosensitizers in Photodynamic Therapy*. In Porphyrin Handbook; Kadish, K. M., Smith, K. M., Guilard, R., Eds.; Academic Press: San Diego, 1999; Vol. VI, Chapter 43. (b) Chou, J. H.; Nalwa, H. S.; Kosal, M. E.; Rakow, N. A.;

Suslick, K. S. *Applications of Porphyrins and Metalloporphyrins to Materials Chemistry*. In *Porphyrin Handbook*; Kadish, K. M., Smith, K. M., Guilard, R., Eds.; Academic Press: San Diego, 1999; Vol. VI, Chapter 41.

(8) Woodward, R. B. *Aromaticity: An International Symposium Sheffield, 1966*; The Chemical Society: London, 1966; Special Publication no. 21.

(9) Broadhurst, M. J.; Grigg, R. *J. C. S. Perkin I*, **1972**, 2111-2116.

(10) Sessler, J. L.; Devis, J. M.; Lynch, V. *J. Org. Chem.* **1998**, *63*, 7062-7065.

(11) Sessler, J. S.; Seidel, D.; Bucher, C.; Lynch, V. *Tetrahedron*, **2001**, *57*, 3743-3752.

(12) (a) Narayanan, S. J.; Sridevi, B.; Chandrashekar, T. K. *Org. Lett.* **1999**, *4*, 587-590. (b) Sridevi, B.; Narayanan, S. J.; Rao, R.; Chandrashekar, T. K. *Inorg. Chem.* **2000**, *39*, 3669-3677. (c) Misra, R.; Kumar, R.; Prabhuraja, V.; Chandrashekar, T. K. *J. Photochemistry and Photobiology A: Chem* **2005**, *175*, 108-117.

(13) Köhler, T.; Hodgson, M. C.; Seidel, D.; Veauthier, J. M.; Meyer, S.; Lynch, V.; Boyd, P. D. W.; Brothers, P. J.; Sessler, J. L. *Chem. Commun.*, **2004**, 1060-1061.

(14) Rao. M. R.; Ravikanth, M. *J. Org. Chem.* **2011**, *76*, 3582-3587.

(15) (a) Holten, D.; Bocian, D. F.; Lindsey, J. S. *Acc. Chem. Res.* **2002**, *35*, 57-69. (b) Shinokubo, H.; Osuka, A. *Chem. Commun.* **2009**, 1011-1021.

(16) (a) Springs, S. L.; Gosztola, D.; Wasielewski, M. R.; Král, V.; Andrievsky, A.; Sessler, J. L. *J. Am. Chem. Soc.* **1999**, *121*, 2281–2289. (b) Král, V.; Springs, S. L.; Sessler, J. L. *J. Am. Chem. Soc.* **1995**, *117*, 8881–8882. (c) Inokuma, Y.; Osuka, A. *Org. Lett.* **2004**, *6*, 3663-3666.

(17) Misra, R.; Kumar, R.; Chandrashekar, T. K.; Suresh, C. H. *Chem. Commun.* **2006**, 4584-4586.

(18) Rao, M. R.; Ravikanth, M. *Eur. J. Org. Chem.* **2011**, 1335-1345.

(19) Pareek, Y.; Ravikanth, M. *Eur. J. Org. Chem.* **2011**, 5390-5399.

(20) Khan, T. K.; Ravikanth, M. *Eur. J. Org. Chem.* **2011**, 7011-7022.

(21) (a) Carrano, C. J.; Tsutsui, M. *J. Coord. Chem.* **1977**, *7*, 79. (b) Mangani, S.; Meyer, E. F.; Cullen, D. L.; Tsutsui, M.; Carrano, C. J. *Inorg. Chem.* **1983**, *22*, 400-404. (c) Marrese, C. A.;

Carrano, C. J. *Inorg. Chem.* **1983**, *22*, 1858-1862. (d) Gouterman, M.; Sayer, P.; Shankland, E.; Smith, J. P. *Inorg. Chem.* **1981**, *20*, 87-92. (e) Sayer, P.; Gouterman, M.; Connell, C. R. *J. Am. Chem. Soc.* **1977**, *99*, 1082-1087.

(22) (a) Ghosh, A.; Ravikanth, M. *Chem. ⁻Eur. J.* **2012**, *18*, 6386-6396. (b) Liang, X.; Mack, J.; Zheng, L.-M.; Shen, Z.; Kobayashi, N. *Inorg. Chem.* **2014**, *53*, 2797-2802. (c) Simkhovich, L.; Mahammed, A.; Goldberg, I.; Gross, Z. *Chem. Eur. J.* **2001**, *7*, 1041−1055. (d) Kadish, K. M.; Ou, Z.; Adamian, V. A.; Guilard, R.; Gros, C. P.; Erben, C.; Will, S.; Vogel, E. *Inorg. Chem.* **2000**, *39*, 5675-5682. (e) Paolesse, R.; Boschi, T.; Licoccia, S.; G. Khoury, R.; M. Smith, K. *Chem. Commun.* **1998**, 1119-1120. (f) Ghosh, A.; Lee, W.-Z.; Ravikanth, M. *Eur. J. Inorg. Chem.* **2012**, 4231-4239. (g) Liang, X.; Mack, J.; Zheng, L.-M.; Shen, Z.; Kobayashi, N. *Inorg. Chem.* **2014**, *53*, 2797-2802.

(23) Młodzianowska, A.; Latos-Graz'yn'ski, L.; Szterenberg, L. *Inorg. Chem.* **2008**, *47*, 6364-6374.

(24) Miura, T.; Higashino, T.; Saito, S.; Osuka, A. *Chem. Eur. J.* **2010**, *16*, 55-59.

(25) Higashino, T.; Lim, J. M.; Miura, T.; Saito, S.; Shin, J.-Y.; Kim, D.; Osuka, A. *Angew. Chem. Int. Ed.* **2010**, *49*, 4950-4954.

(26) Higashino, T.; Osuka, A. *Chem. Sci.* **2012**, *3*, 103-107.

(27) Higashino, T.; Lee, B. S.; Lim, J. M.; Kim, D.; Osuka, A. *Angew. Chem. Int. Ed.* **2012**, *51*, 13105-13108.

(28) (a) Paolesse, R. In The Porphyrin Handbook; Kadish, K. M., Smith, K. M., Guilard, R., Eds.; Academic: San Diego, CA, 2000; Vol. 2, ch. 11, pp 201-232. (b) Erben, C.; Will, S.; Kadish, K. M. In *The Porphyrin Handbook*; Kadish, K. M., Smith, K. M., Guilard, R., Eds.; Academic: San Diego, CA, 2000; Vol. 2, ch. 12, pp 233-300.

(29) (a) Licoccia, S.; Paolesse, R. *Struct. Bonding (Berlin)* **1995**, *84*, 71-134. (b) Sessler, J. L.; Weghorn S. J. In *Expanded, Contracted and Isomeric Porphyrins, Tetrahedron Organic Chemistry Series, Vol.15*; Pergamom, New York, 1997; pp 11-125.

(30) Brückner, C. *J. Chem. Educ.* **2004**, *81*, 1665.

(31) Johnson, A. W.; Kay, I. T. *J. Chem. Soc.* **1965**, 1620-1629.

(32) Harrison, H. R.; Hodder, O. J. R.; Hodgkin, D. C. *J. Chem. Soc. B* **1971**, 640-645.

(33) Johnson, A. W.; Kay, I. T. *Proc. Chem. Soc.* **1961**, 168-169.

(34) (a) Gross, Z.; Galili, N.; Saltsman, I. *Angew. Chem., Int. Ed.* **1999**, *38*, 1427-1429. (b) Gross, Z.; Galili, N.; Simkhovich, L.; Saltsman, I.; Botoshansky, M.; Blaser, D.; Boese, R.; Goldberg, I. *Org. Lett.* **1999**, *1*, 599-602.

(35) Paolesse, R.; Jaquinod, L.; Nurco, D. J.; Mini, S.; Sagoni, F.; Boschi, T.; Smith, K. M. *Chem. Commun.* **1999**, 1307-1308.

(36) Ghosh, A.; Chatterjee, T.; Lee, W.-Z.; Ravikanth, M. *Org. Lett.* **2013**, *15*, 1040-1043.

(37) (a) Broadhurst, M. J.; Grigg, R.; Johnson, A. W. *J. Chem. Soc. D* **1970**, 807. (b) Broadhurst, M. J.; Grigg, R.; Johnson, A. W. *J. Chem. Soc. D* **1969**, 1480. (c) Broadhurst, M. J.; Grigg, R.; Johnson, A. W. *J. Chem. Soc. C* **1969**, 3681.

(38) Broadhurst, M. J.; Grigg, R.; Johnson, A. W. *J. Chem. Soc., Perkin Trans. I* **1972**, 1124.

(39) Lee, C. -H.; Cho, W.-S.; Ka, J.-W.; Kim, H.-J.; Lee, P. H. *Bull. Korean Chem. Soc.* **2000,** *21,* 429-433.

(40) Pawlicki, M.; Latos-Graz'yn'ski, L.; Szterenberg, L. *J. Org. Chem.* **2002**, *67*, 5644-5653.

(41) Shetti, V. S.; Prabhu, U. R.; Ravikanth, M. *J. Org. Chem.* **2010**, *75*, 4172-4182.

(42) Sankar, J.; Anand, V. G.; Venkatraman, S.; Rath, H.; Chandrashekar, T. K. *Org. Lett.* **2002**, *4*, 4233-4235.

(43) Sankar, J.; Rath, H.; PrabhuRaja, V.; Chandrashekar, T. K.; Vittal, J. J. *J. Org. Chem.* **2004**, *69*, 5135-5138.

(44) Sridevi, B.; Jeyaprakash Narayanan, S.; Chandrashekar, T. K.; Englich, U.; Ruhlandt-Senge, K. *Chem.⁻Eur. J.* **2000**, *6*, 2554-2563.

(45) Kim, D.; Osuka, A. *Acc. Chem. Res.* **2004**, *37*, 735 –745.

(46) Osuka, A.; Shimidzu, H. *Angew. Chem. Int. Ed.* **1997**, *36*, 135-137.

(47) Sankar, J.; Rath, H.; Prabhuraja, V.; Gokulnath, S.; Chandrashekar, T. K.; Purohit, C. S.; Verma, S. *Chem.⁻Eur. J.* **2007**, *13*, 105-114.

(48) Mahammed, A.; Giladi, I.; Goldberg, I.; Gross, Z. *Chem.⁻Eur. J.* **2001**, *7*, 4259 –4265.

YOUR KNOWLEDGE HAS VALUE

- We will publish your bachelor's and master's thesis, essays and papers

- Your own eBook and book - sold worldwide in all relevant shops

- Earn money with each sale

Upload your text at www.GRIN.com
and publish for free